This book belongs to:

Remember the Moon

written by Anna Williams

illustrated by Serena Dawson

REMEMBER THE MOON

Author: Anna Williams

Illustrator: Serena Dawson

Design/layout: Niki Merrett

Publication date: 2017

National Library of New Zealand Cataloguing-in-Publication Data.

Williams, Anna.

Remember the Moon/ Anna Williams

ISBN 978-0-473-38626-9

This book is dedicated to the child in all of us.

Thanks to Derin Attwood and my writing group Pen-Ultimate for their ongoing support and special thanks to Wendy Hay for her patient assistance.

What do you see at night,

shining a silvery light?

The moon

What appears each day,

to chase the dark away?

The sun

What moves across the sky,

so very, very high?

The clouds

What blows through the trees,

rustling the leaves?

The wind

What pitter-patters to the ground,

where pretty flowers are found?

The rain

What is soft and cold,

and slippery to hold?

The snow

What do you see at night,

shining a silvery light?

Do you remember ... ?

The moon

You remembered!

The End

About the author:

Anna Williams lives with her husband in Whangarei, Northland. Writing has always been a big part of her life, and while she has had other work published previously to this, 'Remember the Moon' has been a whole new adventure for her and has turned a dream into a reality.

About the illustrator:

Serena Dawson is a writer and award-winning artist who lives on the Tutukaka Coast with her husband and three daughters. She loves animals, the outdoors and sharing the beauty of this world through paint and word. 'Remember the Moon' is the first book she has illustrated.

Made in the USA
Monee, IL
05 November 2026

67386146R00017